Truman Henry Safford

Mathematical Teaching And It's Modern Methods

Truman Henry Safford

Mathematical Teaching And It's Modern Methods

ISBN/EAN: 9783337004576

Printed in Europe, USA, Canada, Australia, Japan

Cover: Foto ©berggeist007 / pixelio.de

More available books at **www.hansebooks.com**

MATHEMATICAL TEACHING

AND ITS

MODERN METHODS.

BY

TRUMAN HENRY SAFFORD, Ph.D.,

FIELD MEMORIAL PROFESSOR OF ASTRONOMY IN
WILLIAMS COLLEGE.

BOSTON:

D. C. HEATH & CO., PUBLISHERS.

1896.

J. S. Cushing & Co., Printers, Boston.

PREFATORY NOTE.

THE following pages were suggested by a long practical experience in giving instruction in mathematical subjects. The conclusions presented in them are believed to be in agreement with the views of progressive educators, and differ essentially from the ordinary traditions. I am only too sensible how imperfectly I have been able to express in words the practice of better teachers than myself, of whom two especially, Professors J. M. Peirce of Harvard University, and George A. Wentworth of Exeter, have given me kind assistance, which I wish here to acknowledge.

T. H. SAFFORD.

WILLIAMSTOWN, Dec. 2, 1886.

INTRODUCTION.

———◆———

THE history of mathematical teaching in this country is yet to be written. It is necessary to pay some attention to this history in writing upon the theory; as the traditions of the elders have a great influence, partly good and partly injurious. If we find that a tradition in mathematical teaching arose from definite reasons still in force, we must be cautious about rejecting it as useless ; but such are not all the methods which have been handed down.

A few salient points of history may be useful to consider. About 1730 the Hollis professorship of mathematics and natural philosophy was founded at Harvard College ; it goes back near enough to Newton's time to enable us to see the impulse which caused its establishment. Again, the successors of Winthrop, the first Hollis professor, taught by English methods and essentially English books, including Playfair's Euclid, till long after the Revolution. Webber's Mathematics, compiled from Hutton and the other English writers, and Hutton's own books, were the standard for many years. It was the military experience of the Revolution which led to the employment of French and Swiss professors in West Point Academy ; and its reorganization under Thayer, shortly after the War of 1812, led to the production by Davies, Church, and Bartlett, of a series of text-books compiled from Lacroix, Legendre, and other writers of the same nation.

But about the same time similar things were done at Harvard College ; and the " Cambridge Mathematics " of John Farrar, similarly compiled, took the place there and in some other newer colleges of the works of Hutton and Playfair.

The marking system, so far as I can gather, seems to have originated in France, — I do not know whether it has yet disappeared there, — and to have been introduced into this country through West Point.

The French mathematics have been the standard originals of our text-books from the early part of this century till the present day. In giving up Euclid, our writers have not been altogether consistent; they have modified Legendre's Geometry so as to make it resemble Euclid as much as possible.

Till within a few years, there have been in this country three main schools for mathematicians, — West Point, Harvard College, and Yale College. Of these West Point was the earliest to educate men in considerable numbers, who could deal successfully with large problems of applied mathematics. The professors, at first partly foreigners, introduced a knowledge of the modern analysis, especially in its applications to engineering and military matters of other kinds; and their pupils were so apt that a single generation replaced the older men with American successors: it was no longer necessary to call men from Europe.

The West Point graduate who has most largely influenced science is doubtless A. D. Bache, a great-grandson of Franklin. The Coast Survey, first conducted by Hassler, had in his hands a character of strangeness; excellent though he was in all his work, he failed to adapt his ideas to some of his subordinates and to certain necessary conditions of American public service; and as he grew older, he found difficulty in carrying his plans out undisturbed. After Bache succeeded to the superintendency of the Coast Survey in 1844, he found available for its higher geodetic works a number of West Point officers, of whom T. J. Lee was one, and Humphreys, afterwards chief engineer of the army, another. One of the leaders in practical astronomy of the Topographical Engineers was J. D. Graham ; and the work which had been done by that corps upon the national and State boundaries had familiarized a good many army officers with field astronomy and geodesy.

Bache, who had been out of the army nearly twenty years, employed his great organizing and scientific capacity in training the Coast Survey corps (including detailed army officers) into practical methods for its various problems ; and the connection between the West Point officers and the able young civilians, who are now the veterans of the Survey, was extremely whole-some.

Lee prepared a work — Tables and Formulæ — which has served an excellent purpose in bridging the gap between theory and practice ; especially for the last generation of West Point officers.

The graduates of this institution, however, are now more closely employed in military and other public duty ; so that their selection as professors of mathematics in colleges is not so common as it was thirty years ago ; and they are no longer employed in the Coast and Geodetic Survey.

The younger naval academy at Annapolis is beginning to send out scientists of ability ; but not in the same line of studies as the military engineers.

Harvard College took up the new movement in mathematics quite early, as I have said. The first great mathematician of Eastern Massachusetts was, however, Nathaniel Bowditch, who was a practical sailor and man of business. His great work — the translation of Laplace's *Mécanique Céleste* — gave a stimulus to mathematics and the higher astronomy which can hardly be overestimated. Among the Harvard men of Farrar's time were James Hayward, Charles Henry Davis, and Sears Cook Walker ; who all in after life did much to promote the science. Hayward was for a while professor, and strove to reform the teaching of geometry ; Davis became a naval officer, served on the Coast Survey, and established the *American Ephemeris:* Walker, after a business life in which he did much for astronomy as an amateur, became also an official of the Coast Survey, and applied the telegraphic method of determining longitudes very early, if he was not its sole inventor. But the great Harvard mathematician was Benjamin Peirce, who made Cambridge a mathematical centre at one time quite unequalled in this country. Peirce was an admirable teacher for those who could receive his ideas; his quickness and depth of intellect were such, however, that he lacked the patience to instruct the average student, and some even of those who wished a high mathematical training were compelled to admit that they could not follow him. He attracted, year by year, a few of the better scholars to his lectures on the Calculus and Analytical Mechanics, and drew around him many of the best mathematicians of the whole country. So that a very large share of those who attended his lectures are now in high positions either as investigators or teachers.

The *American Ephemeris* was in fact established at Cambridge. Peirce and C. H. Davis were nearly connected in family, and worked together in its establishment; and thirty years ago the office attracted young mathematicians by the intricacy of the problems then needing solution, as well as by the very modest recompense which they received. At present the *Ephemeris* is on an excellent footing, and its current work can be done more mechanically: the investigators have not all left it, however.

At Yale College some cause which I cannot precisely indicate — perhaps the influence of Jeremiah Day and Denison Olmsted — gave rise to a school of young mathematicians about half a century ago. Among these, besides men now living, were E. P. Mason, Stanley, — both of whom died comparatively young, — W. A. Norton, but above all, William Chauvenet.

This excellent man and lucid writer was admirably adapted to promote mathematical study in this country. His father, a Frenchman of much culture, trained him very thoroughly in the knowledge of the French language, even in its niceties. They habitually corresponded in that language; and the son was enabled to study the mathematical writings of his ancestral country in a way which enabled him to reproduce in English their ease and grace of style, as well as their matter. In these respects his works are far more attractive than those of ordinary English writers : his Trigonometry is much the best work on the subject which I know in any language; his Spherical and Practical Astronomy is frequently quoted by eminent continental astronomers, and his Geometry has raised the standard of our ordinary text-books, of which it is by far the best existing.

Of late years a new mathematical school — chiefly in certain branches of abstract higher algebra — has been established at the Johns Hopkins University. The original impulse given by Sylvester will, however, be broadened as well as strengthened by the somewhat altered direction lately given to it; and I think that school will produce mathematical physicists as well as pure mathematicians. The great increase of polytechnic schools and schools of science is also favorable to applied mathematics.

It is an old complaint against mathematics as a mental discipline, that it is too abstract and unpractical. When we look at the ordinary courses in our colleges and schools, we shall find that there is much truth in this ; but the complaints are entirely groundless when mathematics takes its proper place in our courses, and is taught in the proper manner.

The theory most prevalent among teachers is that mathematics affords the best training for the reasoning powers ; and this in its traditional form. The modern, and to my mind the true, theory is that mathematics is the abstract form of the natural sciences ; and that it is valuable as a training of the reasoning powers, not because it is abstract, but because it is a representative of actual things.) Thus geometry, as Euclid gives it to us, is an excellent training ; not because the things of which it treats are *unreal*, however : a Euclidian geometry of four dimensions would not be a good study for an ordinary pupil. Again(arithmetic, as we shall see by and by, is over-done, in a certain sense, in our schools ; just so far as the teaching is based upon the concrete, so far is it profitable ; but when the book-makers begin to make it too abstract, as they very often do, it becomes a torture to both teacher and learners, or, at best, a branch of imaginary knowledge uncon-nected with real life.)

The usual requirements for the degree of bachelor of arts in this country include two subjects of rather doubtful benefit to the average college student, — analytical geometry and spheri-cal trigonometry. The justification for the first is complete enough, if by the name a different thing is meant, — conic sections. These must, to some degree, be understood, if physics or astronomy form part of the course ; but their proper treatment for most students is synthetic, although alge-bra may be employed. Of spherical trigonometry, a brief outline may be taught to the average man ; but the real utility of this subject would not compensate him for the pains he would need to take to get a complete idea of it.

In Germany the introduction of both subjects, even into the very thorough gymnasium course, was long debated ; in England neither of these is required for the bachelor's degree at Cam-bridge or Oxford ; at Harvard no one need study any mathematics whatever, even solid geometry ; but the admission examination covers algebra and plane geometry. So that the Harvard mathematical standard for graduation is nominally lower than the German, nominally higher than the English ; while in most other American colleges we profess to require more than do

the institutions of either England or Germany. My own notion is, that the American colleges would do well to require
arithmetic, algebra, up to, but not including, the theory of
equations, plane and solid geometry, and plane trigonometry,
for graduation; and to greatly increase the thoroughness of
teaching within this range. The Yale standard seems to me
about as much too high as the Harvard one is too low, so far
as the amount required is concerned.

A good many years' experience in teaching college students
astronomy, both in Western colleges and at Williams, leads me
to think that the average man has been "over," as the phrase
is, more mathematics than he has digested or can apply. It is
rare that I find such a pupil who is even accurate in arithmetic.
A short course in the art of computation which I give to an
elective class, numbering, on an average, six or eight (introduc-
tory to practical astronomy), is apt to develop great want of
skill in ordinary calculation among the members of the class;
and similar difficulties are apt to manifest themselves in all the
work they have done before entering college. The men who
do not have to be taught very elementary matters in arithmetic, algebra, and geometry, are those who unite natural ability
with a thorough preparation under experienced teachers.
Their college course seems to me to deal largely with too
advanced matters for their real preparation; but it is entirely
the average American course; and I find among the pupils
who come from Western colleges into our upper classes just
the same degree of inability to calculate well and accurately.
In the gymnasia of Germany the teachers are warned not to
suffer ordinary reckoning to be forgotten.

Another quite singular experience has forced itself upon me
again and again, in the most diverse ways. It is that, on the
whole, pupils whose arithmetic was taught them in foreign
countries, especially Germany and the North of Europe, make
fewer mistakes, and are more quickly able to calculate from an

algebraic formula than are those taught in this country. There are German teachers in the United States whom I have known, whose pupils show similar capacity. The reason I take to be this: that in America, arithmetic and the other rudimentary branches of mathematics are taught, not for their own sake, but as a training of the reasoning faculties; and the matter in the text-books is not scientifically arranged, but traditionally; it is not usually selected by mathematicians, but by schoolmasters. These too often require of their pupils, not an independent power of accurate calculation, but the ability to follow and repeat the reasonings of the books, and to solve the various artificial examples — with the answers usually given — which are utterly repugnant, in too many cases, to anything in nature or practical life.

The Hon. J. W. Dickinson, Secretary of the Massachusetts Board of Education, and his agent, Mr. Walton, have expressed themselves very strongly on this matter in the late reports of the Board; the well-known "Norfolk County Report" of Mr. Walton is extremely instructive.

What the cure in detail is for this state of things, will be stated farther on, in treating of the various branches; but in general, it is as follows: —

1. The employment of good, experienced teachers in all schools, and their proper remuneration.

2. A rational course of study, concentrated upon a few essential points, and so graded that the easier matters shall be taught first, and the more difficult ones later. Years ago I found a certain "Higher Arithmetic" more difficult than algebra or trigonometry.

3. The use of small but carefully arranged text-books. Mehler's *Hauptsätze*, the text-book used by many advanced teachers in Germany, gives about 150 pages to algebra, geometry, and trigonometry. This represents an extreme view. I have recommended its reading, in German, to my pupils who intended to become teachers.

 4. Objective and oral teaching, combined with extemporalia upon easy problems.

5. Laboratory work in mathematics, as well as in physics. Every teacher of mathematics should have learned to practically apply the science to surveying and astronomy, as well as to physics. Every high school where pupils are to become teachers, every normal school, as well as every college, should have its modest collection of astronomical instruments. Many high schools in this country are already so provided, and some have completely equipped observatories, — the Hartford (Conn.) High School, for example ; Mount Holyoke Female Seminary, which is almost a college, however ; as well as, in England, Rugby School ; whose astronomical teacher, Wilson, became head master of Clifton College.

Elementary schools should teach carefully ordinary methods of weighing and measuring, both on the common and on the metric system.

The qualifications of the teacher or professor of mathematics should, on the mathematical side, include two things. First, a knowledge of more subjects than he expects to teach, always a grade higher. That is, whoever is to teach arithmetic must know algebra, including logarithms, and geometry ; the teacher of plane geometry must be well versed in solid geometry and plane trigonometry. Second, a knowledge of the applications of the subjects to be taught ; thus the teacher of spherical trigonometry must be familiar with practical astronomy and geodesy ; and he who only teaches plane trigonometry must know ordinary surveying. The teacher of arithmetic in a common school must be thoroughly familiar with weighing and measuring, and with the practical application of the metric system.

METHODS OF TEACHING THE VARIOUS BRANCHES OF MATHEMATICS.

———◆———

I SHALL take them up in the following order : Primary Arithmetic; Scientific Arithmetic and Algebra; Geometry; Plane Trigonometry; Algebraic Geometry, Analytical Geometry, and Descriptive Geometry; Spherical Trigonometry; Differential and Integral Calculus, Higher Analysis.

This order is not that of a well-arranged programme of study, but is (approximately) according to the usual practice. The right position of the separate branches will be indicated later. Certain general remarks can, however, be made in this chapter which refer more or less to all mathematical teaching, from that of the simplest numbers up to that of the highest analysis.

The first of these is, that all teaching should have a real, objective basis.

To show what this should be, I will take a moderately easy example from elementary algebra; logarithms. The theorem on which this depends is the following :

$$a^m a^n = a^{m+n}$$

This is true whether m and n be positive or negative, integral or fractional. The pupil is supposed to know this theoretically and practically, root and branch; and to have applied it to a sufficient number of easy cases in which, for example, the denominator of m and n shall not exceed 10, to be thoroughly convinced of its truth and practical applicability; and to understand the logical limitations of the formula so far as not to confuse different roots of the same number. In other words, he must know that the number a is positive, and its positive roots only are to be considered.

The table of logarithms is now the objective basis. The teacher explains that $a = 10$ and that the values of m corresponding to various values of a^m are given in it. The pupil will naturally ask how this table is formed; and the teacher, *if competent*, can give a sufficient answer. After this is done, multiplication by logarithms and division by logarithms are now practised until the class is thoroughly trained in these two rules as far as they can be employed without interpolation.

The omission of the characteristic in the table has to be explained by the teacher during this exercise, which is not completed until it is thoroughly understood, and, in fact, becomes mechanical.

The number of decimals employed in the pupils' tables should be five; four or six will do, however. The library of the institution should have at least four tables accessible, — Bruhns's seven-figure tables, Albrecht's edition of the six-place Bremiker, and some good five and four place tables. The two just mentioned are easily obtained, and not expensive.

Interpolation and the limits of accuracy in the use of the tables should now be explained and made secure by practice.

After multiplication and division by logarithms have been thus taught, the subjects of involution and evolution should be treated in the same manner. Those of the class who are likely to become teachers must here be shown that the practical method of dealing with powers and roots is the logarithmic; so that in later years they will abstain from annoying their young pupils with difficult and needless problems solved in the antiquated manner. And they will be shown, also, how to calculate a compound-interest table, an excellent exercise in itself, as well as a labor-saving contrivance in arithmetic. In dealing with roots and compound interest, interpolation must be rationally explained, and not blindly executed. If, now, we ask why logarithms are so little appreciated in business and engineering, we shall find the reason to be that the teachers of arithmetic have not, as a rule, really learned their use; they have "been over the subject," it is true, so that they go on wasting time in arbitrary exercises in involution, evolution, and compound interest done by more tedious methods, and do not really appreciate how instinctively the best calculators employ logarithms.

Objectivity in teaching them consists, then, in making logarithms the object. But to give a good object-lesson of any kind, the teacher must know his object not merely superficially, but thoroughly, and with interest in it. The man or woman who is to deal with logarithms as a teacher, must be in some degree a practical computer.

Let us take another example from a subject which is often badly taught, — solid geometry.

The objective basis here is a perception of the relations of space. It is a difficult thing to produce, and that pupil is fortunate who has had really good object-lessons on form at and from an early age.

Solid geometry is usually taught to Freshmen in college. If the instructor has himself a vivid perception of form, he can by taking pains and time, and exerting himself, do a good deal with his men at the beginning of the subject, by employing quite common illustrations, such as the walls of the class-room, the doors, the outside walls of the building, etc. In small classes the stereoscope can be used to advantage; there are stereoscopic views of geometrical solids to be had in the market; and a little pains will enable him to employ the magic lantern. In all his teaching he must not forget that his end in view is to produce images of the geometrical figures in the minds of the pupils; so that he and they will be looking mentally at the same or similar objects; and that neither will be lost among the abstract words of the demonstrations. Any other teaching of solid geometry is a mockery, and had better be replaced by something more useful.

A third example of objectivity in mathematical teaching is the differential calculus in its rudiments. The object is best expressed in Newton's words, — the fluxion and the fluent. When the great philosopher discovered the calculus, he simply generalized certain principles which had preceded it; and to do this was obliged to concentrate his thoughts upon the objects just named, not as *functions* of anything else, nor as representatives of any mechanical process : but as variable quantity and its variation.

The illustrations which naturally arise in a teacher's mind, and the definitions and boundaries which have to be set up in teaching, are not the principal matter ; the concepts themselves of function and differential coefficient are the primary objects of contemplation.

The pupil's mind does not grasp the elements of the calculus until these primary objects are wholly present to consciousness, and he can practically and intelligently apply these conceptions.

In a far lower degree the same method is the true one for the elements of number. The objects here are at first the *individual numbers* as Grube has pointed out ; and when these are mastered up to 100, the larger concept of number in general takes their place. The analysis of the number 4, or the number 11, or the number 97, is as difficult to one pupil as that of the concept " function " to another.

The grand stumbling-block of all inexperienced teachers is dogmatic method. Just as the old Eton Greek and Latin grammars — written both of them in Latin — were incomprehensible to the pupil, so are the definitions of any branch of mathematics to the pupil utterly unfamiliar with the subject. It may be necessary, and it certainly is unobjectionable, for the text-book to express the rudiments of knowledge on any subject in a few pages. But the teacher who gives out these few pages as a lesson to a pupil ignorant of the subject has not yet mastered the alphabet of his profession. It is his duty to commence by a careful study, with a mental sounding-line, of the depths of his pupils' ignorance ; to be followed by an objective presentation of the elements of the new subject. There are various ways of exciting curiosity ; in mathematics it is unusually necessary to employ them, as most scholars come to the higher branches a good deal prejudiced against them. Poor teaching leads to the inveterate idea that the subject is only adapted to peculiar minds, when it is the one universal science, and the one whose four ground-rules are taught us almost in infancy, and reappear in the motions of the universe.

In fact, I cannot foresee the extent of the beneficent effect of genuinely good teaching (viz., according to Grube's method) of primary arithmetic. It seems to me almost as if it would revolutionize the whole mathematics up to quaternions ; at least, if combined with Pestalozzian teaching of geometry *and drawing*.

PRIMARY ARITHMETIC.

The space is here lacking to set forth Grube's method in detail.* Excellent pamphlets have been published upon it ; the one written by Prof. F. Louis Soldan is the most available which I have found in English.

I am inclined to think that the strict practical application of the first principle — analysis of *every* number up to 100 — requires more arithmetical power than the ordinary untrained teacher possesses. For such a one, then, to attain what I consider the highest success, it is necessary to work out

* See Appendix.

this analysis carefully in detail, and impress it upon the memory.   Let me give an example — the number 89 : —

$$89 = \text{a prime number}$$
$$= 11 \times 8 + 1$$
$$= 29 \times 3 + 2$$
$$= 43 \times 2 + 3$$
$$= 17 \times 5 + 4$$
$$= 7 \times 3 \times 2^2 + 5$$
$$= 83 + 6$$
$$= 41 \times 2 + 7$$
$$= 3^4 + 2^3 ; \text{ and so forth.}$$

In other words, the teaching of any *one* number of the hundred involves on the teacher's part an *immediately available* perception of every combination of factors in every number less than that one ; as well as a ready memory of all combinations of two numbers which produce it by addition.   Grube does well in restricting to 100, the numbers to be *individually* analyzed ; does he put too severe a strain upon the teacher in requiring this method up to 100?   Certainly not, if the teacher is selected for his or her skill ; very possibly, if the ground of election is poverty, friendship, or any other quality not available to the good of the pupils.

Is Grube wrong, then, in thinking it possible to accomplish the work up to 100 in two years, beginning at the age of six or seven?

Or shall we say that so thorough a knowledge of these numbers is not worth communicating?   It seems to me almost self-evident that it is, — if arithmetic has any usefulness.   For the lack of skill *within these limits* among business men is almost inconceivable, to say nothing of those who are dependent upon their daily labor.

We may, if we choose, take the ground that thorough teaching of any kind is impracticable, theoretical, visionary ; that teachers are entitled to do bad work because their salaries are so poor; or that the instincts of a young person of either sex are a better guide in this profession than the thoughts of a Pestalozzi, or a Froebel, or their followers.

But the present writer can assure all who employ arithmetic that the thorough analysis, strictly according to Grube, of the first hundred numbers, both concrete and abstract, is a most admirable foundation for higher arithmetic, and a most useful exercise preparatory to every-day business.

Who does not sometimes wish to know *what he can do* with a definite sum of money? Is not, in fact, this the really fundamental problem of business arithmetic?

In all rational methods of teaching arithmetic, the mental or oral side of the subject is the most important. Here, again, the inexperienced teacher is at a loss. It is comparatively easy to set sums from the book, and see if the answer agrees; and but little more difficult to consult the "key"; not so easy, however, to make up, and control the solution of, oral extemporaneous examples.

It is here that the normal schools afford the opportunity of professional training, too precious to be lost. Even college professors could, if they would, do some good in the same direction; there are many chances for them to teach the importance of mental arithmetic.

Written arithmetic should not be taught with large numbers. The pupil who can add, up to five significant figures, rapidly and correctly; or multiply two numbers of four digits each by each other; or divide eight figures, or even six, by three, without frequent failure, is well enough equipped. It is not in this direction that practical weakness in arithmetic lies.

Fractions, as taught according to Grube, are first objectively presented. The addition, subtraction, multiplication, and division of the simpler ones should precede any dealing with more complicated forms. Thoroughness in those which — under any circumstances whatever — are oftenest used is the main thing to be aimed at, after all.

It will be found by any one who notices the calculations made by common people that halves and quarters are very much oftener used than thirds, sixths, or even tenths. The carpenters' rule gives eighths and sixteenths of an inch; the mechanic is a very fine one who employs decimal fractions at all. Federal money is, practically, dollars and cents, hardly calculated as dollars and hundredths except in banks or other large business.

Thus in teaching fractions the simpler ones should be utilized as far as they will go, and those more complicated than any one uses left out of view entirely. There is a justification in the employment of all denominators up to ten, but for few primes beyond ten save very incidentally.

It is a common practice in modern text-books to introduce decimal fractions before vulgar. I suspect it not to be the natural order; those who so present the subject probably expect the pupils to learn the simpler vulgar fractions earlier, perhaps in an easier book. Denominate numbers are (usually) overdone in our arithmetics and in those schools where the book, the whole book, and nothing but the book, furnishes the material of teaching. It is of some use to a surveyor to know the number of feet in a mile, and perhaps the number of inches; but the dealers in square feet of land, and those in square miles, are too far apart to have any important relationship. There are absurd questions of reduction ascending and descending which are only a torture, and I doubt the possibility on the part of ordinary people of reading an ordinary physician's prescription so as to apply apothecaries' weight. We are more liable to be poisoned by the wrong substance than the wrong quantity; if the doctor and the apothecary can learn to read the dog-Latin, they can also get up the weights and measures which accompany it. And I fancy the homœopaths employ a more rational system of measurement, however great be the strain they may be thought to put on the powers of numbers.

Percentage and its applications, except interest, are a pretty easy subject; and interest is actually taken up in our schools and our books far beyond the necessities of life. Partial payments, it is true, are sometimes troublesome, but experience shows that the power of doing such examples does not follow the study of ordinary arithmetic unless the pupil has learned (as few have) to do simpler sums correctly. The writer was called upon a few years ago to do such a calculation, which had been "given up" by the holder of the note, a large merchant in a great city; by the maker of the note, of course (he was only an intelligent native of a foreign land); and by his two daughters, normal school graduates, and actual high school teachers in Massachusetts. It was no more difficult than many given in our books of arithmetic, *but none of, the parties knew the answer.*

Compound interest should be lightly touched upon; the table of amounts for various percentages and times is easily proved and understood, and is very instructive to the extravagant and the improvident.

Proportion is best taught in the beginning by the "reduction to unity"; in fact, it is better called rule of three at first, and when a little more abstraction begins, the use of letters for numbers should begin also.

Can any one imagine a good teacher who is also a good algebraist, who will not train his pupils to use letters for numbers long before arithmetic is completed? Interest and proportion are the two most important portions of arithmetic for this training; but until the present writer actually takes a class through the whole course from eight or ten years of age to fourteen, he cannot say positively when he would begin the use of letters. They would certainly be in place long before the time indicated by the present custom.

Mensuration is the application of arithmetic to geometry. The arithmetical part is simple enough; but the geometry must precede it, and be well taught, or that will happen which was told me, that only one man in a village (not a college professor) could measure a wall correctly. The one man became extremely powerful in politics. I do not say that the "boss system" in our politics is due to ignorance of mathematics; but the arithmetical power of a leading politician must evidently be considerable.

In conclusion of the chapter, let me say that a good teacher of arithmetic must combine the following qualities : —

1. Quickness in mental operations.

2. Correctness in calculation.

3. Power rapidly to form new examples, especially in concrete numbers.

4. Knowledge of algebra and geometry.

5. Ability to teach objectively and find illustrations.

6. Patience with the slow pupils.

7. Thoroughness everywhere.

To improve in teaching arithmetic, he or she must improve in all these qualities; the best books on the advanced branches, so far as I know them, will be mentioned later.

SCIENTIFIC ARITHMETIC AND ALGEBRA.

We may, if we choose, make a distinction between higher arithmetic and elementary arithmetic on the one hand, and algebra on the other. This course is conducive to that multiplication of books which is so dear to some authors and some teachers. But a practical distinction of great importance is between the arithmetic which every man — even a day-laborer — needs, and that which is more technically business arithmetic. The day-laborer needs to have a very thorough acquaintance with the four ground-rules and federal money, in order to escape the snares which are set for his ignorance ; and the common school cannot altogether restrict its course to this amount. A knowledge of the intricacies of business arithmetic is best deferred till the young man needs to study them professionally.

Now my contention is, that after elementary arithmetic through fractions, rule of three, and percentage, and such a course of these subjects as shall deal only with moderate numbers and involve much mental arithmetic, the next subject on the arithmetical side should be elementary algebra.

But this should be first taught in a *propædeutic* way ; that is, orally and objectively, in 'the proper manner to introduce a new subject. The average age of the class beginning algebra so will be about *thirteen*. The whole year's work in algebra will be given to the four ground-rules, with a few simple equations, to excite more interest ; and negative quantities will be taught after these have been dealt with, employing positive quantities, added and subtracted. That is, in the early exercises care will be taken that the difficulty of the subtraction of a larger number from a smaller does not arise.

The method of treatment can be got from the first five chapters of Todhunter's Algebra, with the application of a little common sense in the introduction of abstract terms and ideas.

But, above everything else, let the teacher practise his or her class patiently, diligently, and faithfully, both orally and in writing, with simple examples rather than complicated ones, with lively dialogue rather than tedious recitation, on every point involved. Two lessons a week for a year can be well spent on these chapters.

Square and cube root applied to numbers, and compound interest, should be taught in this year.   ·

Let the pupil who is to be educated for commercial business

learn a good deal of algebra thoroughly before he begins his technical course.

After the four ground-rules of algebra are really well taught, the subject is very easy until we come to radicals. The combination of exponents is very important and necessary, easy enough so long as the exponents are whole numbers, but a little difficult when they are fractional or negative. Their difficulty can be avoided, first of all, by objective, oral, practical treatment; but is greatly aggravated by weakness of the pupil in the four ground-rules, the use of the bracket and negative quantities included.

How many become discouraged at this point! Our colleges admit many Freshmen whose ideas of radicals are utterly confused, who, in consequence, cannot do anything with quadratics.

Let me say here that the teacher of algebra who does not keep his pupils fresh in ordinary arithmetic by continually reviewing the important portions — setting problems orally — loses a great aid in mathematical training.

Let me take a brief diversion here. The practical mathematician, I do not care what his profession may be, spends a large share of his time in the ordinary operations of arithmetic, simple algebra, easy geometry; and the time given to difficult and perplexing problems of the higher mathematics is small indeed. If he be one of those rare men who can solve the great problems of abstract mathematics, he will soon find himself obliged to live alone, with hardly any sympathy; and the few disciples he can get will, like himself, be in danger of becoming hopelessly unpractical. This will happen unless he takes a lively interest in some branch of natural science proper.

Whoever you be, and whatever branch of mathematics you wish to teach, do not forget to practise your pupils upon the elements as well as upon the higher branches. Just as soon as the theory of combination of exponents has been taught, both for whole numbers and fractions, logarithms should be introduced and practically employed. They should certainly precede quadratic equations.

Algebra, "through quadratics," is in Eastern colleges the usual requirement for admission. In this requirement logarithms are not included. The usual result is that these are not practically taught until trigonometry is begun, and the average pupil does not learn the difference between a logarithmic sine and a natural sine. There is no cure for this confusion, except to teach logarithms where they belong, — as a part of scientific arithmetic, — and to apply them to purely arithmetical problems, and to be thorough with their practice.

The binomial theorem is often taught by the method of indeterminate coefficients, very loosely applied. This is a hasty and untrustworthy method; the *demonstration* in this manner requires several steps usually overlooked. In this, as in many other cases, I think the usual fashion of teaching goes too far. A boy is better prepared to take up the differential calculus, if he knows the rigid demonstration of the binomial theorem for positive whole exponents only, and can apply it rapidly and accurately, than if he has only a superficial acquaintance with it by the method of indeterminate coefficients. Of easily accessible works upon algebra I should recommend Todhunter as the best work for a teacher to study. It is not altogether easy to employ it as a text-book in the ordinary schools of this country, for the reason that it is pretty large, and has more adaptation to English methods of instruction. It was originally designed for those students in Cambridge, England, who were aiming at mathematical honors, generally under private instruction as well as public.

The theory of equations is frequently studied in our colleges. It should properly be replaced (as a required study) by more thorough training in quadratics and cubic equations, using in the latter Cardan's rule for reducible cases, and the trigonometrical solution for the irreducible. In point of fact, it is hardly worth while to take up the theory of equations at all until the pupil has become thoroughly familiar with trigono-metrical analysis, so as to be able to handle complex numbers with facility.

The old system of college tutorships, now in its decadence, is responsible for a great deal of harm in mathematics as well as in other subjects. The young man just out of Harvard was often employed as a tutor, while pursuing law or divinity as a student : the now most eminent English scholar in the country, being valedictorian of his year, was selected as such a tutor in *mathematics* a year after his graduation. Of course he went through the form perfectly ; but his heart cannot but have been elsewhere. Within a few years this has been changed in nearly all good colleges. Men are selected to teach college students with some reference to their own tastes and studies ; and the corresponding requirement that those studies shall be kept up is more insisted upon.

These remarks are made at this point because the attempt to teach higher algebra in the Freshman year is often the most conspicuous failure in our mathematical courses. The young men visibly love to *practise* that which is not too difficult. A skilful algebraist who is also an *experienced* teacher can make the subject a delight to many, if he is not, as sometimes happens, simply inactive and critical. But to make it a delight to any one, that pupil must be doing something well within his powers, and not be attempting to aim blindly at what is far beyond his capacity.

The good time, however, is coming when temporary teachers of all grades will, little by little, give place to those who intend to make teaching a permanent profession.

To conclude what I have said upon algebra, let me give the following rules for teaching it : —

1. Base it upon arithmetic, and keep up a continual set of exercises in mental arithmetic and mental algebra.

2. Use extemporalia very frequently.

3. Keep yourself fresh in the subject.

4. Learn the secret of *objective* teaching even in the most abstract matter.

5. Be not afraid of too much practice or too much repetition of elements ; look at the way your pupils practise ball-playing.

6. If you are compelled to use a bad text-book, supply its deficiencies by much study of good ones ; and remember that the text-book is, after all, not the teacher.

7. If you are an inexperienced teacher, get advice from older and wiser heads ; do not think that when you have learned the subject, — perhaps in the antiquated manner, — you can teach it without experience or caution.

8. Study the writings of advanced educators.

GEOMETRY.

There is nowhere in teaching any greater gap between theory and tradition than in this subject. It is quite possible to obtain what are considered good results in geometry by the ordinary method ; viz., that of requiring the reproduction by the help of the blackboard of the demonstrations in the text-book. The pupils taught in this way, however, are usually unable to apply geometry, and especially solid geometry, to any advantage ; and when, if it so happens, they are required to do this, they are obliged to learn over again the rudiments, which in all probability they have slurred over or neglected at the proper time. The process is like that of building a foundation for a house already completed, which is very proper in the case of ready-made houses ; but the final result is not altogether the ideal one, nor will the house so constructed be permanently satisfactory.

Dr. Thomas Hill, ex-president of Harvard College, insists very strongly upon the early presentation of the facts of geometry. The same idea has been carried out in foreign countries quite extensively these many years, — it was one of Pestalozzi's, —but does not yet meet with very great acceptance in America. The reason I do not know, except, it may be, that the traditional course is quite different, and that school organization in this country is often, in fact usually, in the hands of persons who are not educators ; and that even the best teachers are allowed but little freedom.

The kindergarten, as I have said elsewhere, introduces forms in the concrete ; Pestalozzi and Pestalozzians make much of forms in the primary school ; but in our practice this knowledge is forgotten before geometry is nominally begun. The contention, then, of theory, is to introduce geometry before arithmetic is completed ; and I am very much inclined to think that this should be done more thoroughly and skilfully than is often the case, even if some arithmetical subjects need to be displaced to make room for it.

The method of beginning geometry while arithmetic is going on is of course the objective one. Let the teacher present for analysis a cube of wood ; call upon the pupils to count its faces, edges, corners ; do the same with the other regular solids ; afterwards, with some not regular polyedrons, which are of practical importance, and with the round bodies. Meanwhile, let the idea of a geometrical surface — a line, a point — be abstracted from the concrete presentation ; and after the concept of the geometrical point has been thus obtained, let the teacher proceed to form the different lines — straight, broken, curved — by the motion of a point, the different surfaces by the motion of the lines. Experience must show in each case how far to go.

For a class which is to study mensuration, the few theorems necessary will offer no great difficulty. Strict demonstration is not needed here; it is sufficient if the pupils are taught in a semi-objective way the rational groundwork of the truths, whose complete comprehension is the main thing needed. Even the Pythagorean proposition can be illustrated by several methods, none of which are equal to Euclid's, in my judgment, as demonstrations; for when you try to put them into words, you find more words are needed; but one or two at least are better for the elementary, half-demonstrative stage preliminary to mensuration.

For an older class which is beginning geometry with the intention of going through an ordinary course in Euclid or Legendre or the late writers, but of which none or few of the members have any good preparation of geometrical ideas, the teacher's difficulties are at the maximum. Even here something may be done by giving a number of lessons on the definitions, and dwelling longer than usual on the first book. The definitions must be objectively illustrated; the same course, of abstracting from the concrete, and of forming straight lines and curves by the motion of a point, must be fully pursued; although as the pupils are older and riper, this will not take so long. In the subject of quadrilaterals a good deal of interest can be roused by a blackboard analysis of the various cases of combinations of four lines; the pupil should be shown the combinations of three lines also, but this is almost too easy.

Three lines can be all parallel.
> two parallels intersected by a third.
> no two parallel, — the triangle.

The first book of geometry, either Euclid or Legendre or Chauvenet, contains enough matter to occupy a year's time. Not five hours a week probably; but the mental growth of a year is not greater than the step from ignorance of demonstration to a thorough mastery of it even in a limited matter. Let me give the course laid down by Arnold (of Rugby) in his

Miscellaneous Works (N.Y., 1845, 8vo) in the matter of
Euclid : —

Fourth Form :   Book 1, Propositions 1–15.
Upper Remove : Book 1, Propositions 15 to the **end.**
Lower Fifth :   Book 3.
Fifth Form :   Euclid to the end of Book 6.
Sixth Form :   Euclid, 3–6; Plane Trigonometry; Conic Sections.

These five classes represent nearly a year apiece. But after
the first book is accomplished, plane geometry has but few difficulties. The English do not learn the quadrature of the circle ; our boys do. It involves pretty long calculations, which
can be abridged, if necessary, by omitting the decimals beyond the fifth place, and made more intelligible by introducing
algebraic notation. This comes in just as well here as later in
geometry.

I strongly advise keeping a small corner of time for the
"modern geometry." Chauvenet's treatise has a small amount of
this in the text, and more in the appendix ; the last editions of
Loomis give a brief appendix to the subject; Gillespie's Surveying introduces it as a help to practical field methods ; and
in several ways the teacher can get possession of some of the
most important results. The foreign text-books already introduce more of these theorems into their main framework.
Steiner, the great Swiss geometer, was Pestalozzi's pupil at
Yverdun ; and his work grew directly out of Pestalozzi's ideas.

It is very likely that the teachers of geometry in our high
schools are often so overworked that they have not much
time to study the many new books on this subject. Riders, as
they are called in England, or "original propositions," as we
call them here, are quite indispensable.

Solid geometry is firmly established in our higher courses ; it
is very singular that in England it is but slowly introduced.
We have excellent text-books, but I suspect the teaching is

not always excellent. Objective illustration, as I have before stated, is the main requirement of the teacher; and he must insist thoroughly and patiently on applications in mensuration. If the little "modern geometry" hinted at above be at hand, the forms in space will take upon themselves a new attraction. Allusions (at least) to perspective, should be made.

Conic sections, in their synthetic form, should precede analytical geometry. The old "Bridge" and his imitators are better than no teaching; but it is to be hoped that the method of Chasles and Steiner will soon become available. If the teacher has ability, the most important properties of the conic sections can be taught *heuristically* by this method, without text-book. It is a very good experiment to try on a strong class of Sophomores.

The teacher's reference-books in geometry may include Chauvenet's Geometry (published by Lippincott), MacDowell's Exercises in Euclid and Modern Geometry (London: Bell), Cremona's Elements of Projective Geometry (Oxford: Clarendon Press; N. Y.: Macmillan), and the works of Chasles and Steiner, if he understands French and German: Mulcahy's Modern Geometry (Dublin) is also excellent.

PLANE TRIGONOMETRY.

If I have not sufficiently insisted upon measurement before, let me now make amends. Before the pupil takes up this subject, he or she must have a distinct conception of the measurement of angles. Paper protractors are good; a pocket compass is of great help; but I think every high school ought to have a theodolite. If nothing more can be procured, a little German instrument with four-inch circles, and a telescope magnifying ten times, can be bought for $60 or $70. Let the teacher lay out and measure the sides and angles of some triangles, and get the pupils to repeat the measures. If a compass only can be employed, the accuracy will not be very great; but still some approximation can be obtained by frequent repetition of the same measures.

The subject of logarithms must be introduced in algebra pretty thoroughly, as I have before stated. The *first* trigonometrical calculations, however, *must* be made with the natural sines, tangents, etc. It is necessary to teach a few formulæ so before proceeding to the logarithmic tables.

Here, as everywhere, the objective method of teaching must prevail. It is pretty easy to show by the circle what the sine, cosine, tangent, cotangent, secant, and cosecant are (of course to radius *unity*). But it is best to introduce at first the sine and cosine only, and show immediately how their introduction reduces the solution of right-angled triangles to order. Now the table of natural sines and cosines is brought forward, and its formation explained in an elementary way; and it is then used in numerical practice. It can also be shown how to plot angles by the help of this table, and they can be compared with the same angles as plotted by a protractor.

Then may be similarly introduced the tangent and the rest of the trigonometric functions, and the solution of right-angled triangles pretty thoroughly mastered.

It is now time to explain the tables of logarithmic sines, etc.; and the pupils are trained in their use with right-angled and oblique-angled triangles, the formulæ for the latter being developed in their order.

The applications of plane trigonometry to surveying are extremely important. Our American surveying methods, as handed down by tradition, have been greatly improved within a few years. The teacher of trigonometry ought to have for reference the latest books; the best are W. M. Gillespie's Surveying, his posthumous treatise on Levelling, Topography, and Higher Surveying; and, above all, J. B. Johnson's Surveying.

It seems to me proper to teach the practical side of trigonometry first; the theoretical part, analytical trigonometry or "goniometry," should be made a strong mathematical discipline, if possible; otherwise not meddled with. The capacity to measure and solve triangles is of practical value; an imperfect knowledge of goniometry is not useful, but rather tends to confuse and dishearten the pupils. In this subject, as everywhere else, teach the concrete first, thoroughly; lead up to the abstract by easy stages.   .

ALGEBRAIC GEOMETRY; ANALYTICAL GEOMETRY.

By *algebraic geometry*, as distinguished from *analytical geometry*, I mean, of course, the application of algebra to geometric problems. It has nothing to do with loci, which are best taught in plane geometry; nor with co-ordinates, which come later. Most treatises on analytical geometry have a short introduction of algebraic geometry; this should be taken by the teacher as the basis of oral lessons, accompanied with extemporalia, and other enlivening exercises. I have already stated that algebra ought, in my judgment, to be pretty freely used in pure geometry, *after* the pupil has been thoroughly trained in geometrical reasoning.

When algebraic geometry has been well taught, and the conception of a locus is well established, there is no difficulty in explaining rectangular co-ordinates. In my judgment this subject should be begun in a very elementary way, and applied first to the ellipse. The straight line and circle are too familiar for the student to see the necessity and meaning of the method; all the more as the properties of the conic sections are practically more important than the abstract method.

When the pupil has a pretty clear idea of rectangular co-ordinates, and of the commoner properties of conics, then, I think, is the time to begin the higher generalizations.

In technical schools a course of descriptive geometry is necessary; and it is much to be wished that it might be brought in as an elective in colleges, as at Harvard, and other colleges which have scientific foundations annexed. The difficulties are pretty serious arising from the general lack of training in drawing.

SPHERICAL TRIGONOMETRY.

Spherical trigonometry is a branch of the science which at Harvard and in all English courses is quite optional. It is rather remarkable that it is not required for all pupils at the Massachusetts Institute of Technology. A certain amount of this branch is needful for a complete understanding of solid geometry; but I do not think that the subject is well taught in many institutions. In fact, if astronomy ever gets its rights in our courses, a rational disposition can be made of spherical trigonometry, which is not now possible.

Its principal uses are

    1st. In crystallography.
    2d. In the theory of surveying-instruments.
    3d. In geodesy.
    4th. In astronomy.

In each of these branches — and they are all pretty advanced subjects — the form of treatment varies with the use. Such a book as Chauvenet's, which is very complete, gives too much for a required course; if the plane trigonometry be taught from this text-book, a slight taste of the spherical can be added. For an elective course in spherical trigonometry or any of its applications, however, the better preparation is a thorough knowledge of plane trigonometry, especially ability to calculate from a formula, with or without logarithms. A smattering of spherical trigonometry, especially when Napier's Rules are the mainstay, is comparatively of no use whatever. Of course spherical trigonometry must be preceded by a thorough course in goniometry.

In teaching an elective class in practical astronomy I find it necessary to begin with a short course in the art of computation. The pupils are ordinarily familiar with common six-figure tables; the first thing to be done is to teach them the use of the modern standard six-place tables (Bremiker's), and to practise them in the computations of spherical trigonometry, employing the fundamental formulæ with partial omission of the logarithmic modifications; also in computations by Gauss's formulæ; in checking their work by various cross-controls; in the use of the modern methods of calculating small arcs; in the calculation of geodetic or small spherical triangles; in the determination of spherical excess and the computation of geodetic triangles by plane trigonometry; and in the use of sum and difference logarithms. Parallel with this course they are taught by daylight the handling of the smaller portable instruments; one of which is a theodolite (or universal instrument) with microscopic readings; and the delicate striding level of this instrument is tested with a level-trier. Thus from a good deal of practice in exact angular measurement (to seconds) they are led to see the definite relationship between observation and calculation, which is very important if spherical trigonometry is to have any special educational value.

THE CALCULUS.

In very few large colleges at the present day the attempt is made to teach the calculus as a required study. The great universities of this and other countries give it a very prominent place as an elective; and in technical schools it is of course required. In some " Realschulen " preparatory to polytechnic

institutions it is required, and with success ; in others a preparation only is made for its future study.

In this country the subject is at a disadvantage from the
want of thorough preparation for it ; the students have too
often slurred over their algebra and trigonometrical analysis.
Granted a thorough knowledge of these subjects, so that the
pupil can quickly follow a well-graded text-book or a judicious
lecture, and progress in the calculus will be relatively rapid.
But at best the instructor must go slowly over the fundamentals,
putting in extemporalia and oral instruction of all kinds, and
carefully making his ground secure as he goes along, taking
care also to guard his pupils from confusion as to the source of
their difficulties.   In the use of a good modern text-book with
examples, the students' greatest difficulty is at first to reduce
the examples to the form indicated by the text-book, or to any
simple form whatever.   This is not a difficulty in the calculus
itself ; for the teacher's business is to explain such matters at
first, until the men are stronger in the subject.   But the most
disheartening thing in mathematical teaching is to find that the
pupils cannot do the elementary part of their work because of
hurried and superficial study (of algebra or trigonometrical
analysis when the calculus is the advanced subject) upon the
lower necessary branches.

It is hardly needful to give directions for instruction in the
portions of mathematics which lie beyond the differential calculus.   There are few who study them ; and the teacher will
almost always do better if he employs the genetic method, or
that which presents the matter in the natural order.   For
example, the theory of determinants is a very important subject ;
while its fundamental idea is an elementary one, the pupil
cannot go very far without some facility in mathematical thinking.   The first notions are communicated with great ease ; but
the study soon leads into pretty deep water.   The teacher will
succeed best with it if he follow in some degree the historical
order of ideas.

Higher algebra, including the theory of equations, and above all, complex numbers, is a branch which should run parallel to the calculus for a good while ; but not much should be done with it before the rudiments of the calculus have been studied. I apprehend that in the foreign universities as well as our own the instinct of the younger mathematical generation is a pretty safe guide as to questions of order. It is only in the schools where the pupil is compelled to follow the traditional programme that there is much danger of intermingling very easy and very difficult matter.

The integral calculus may be divided into several portions. The elementary methods of integration are easy enough, and the pupil can be readily practised in them. But the instructor, and especially the text-book maker, is liable to confuse easy with difficult problems in a very perplexing way. I quite well remember the hopeless dismay with which I read certain books on the integral calculus, only to find that, as I became more familiar with the differential, the difficulties seemed to lessen so far as I had time or need to follow them up. This method left me with vast tracts of the higher portions about which I did not know or care anything until the time came up to employ them, when they appeared easier. In other words, the really important mathematics for practical purposes always grow out of the uses which they subserve.

The university teaching of pure mathematics tends always to take the form of instruction from a discoverer to his disciples. This is proper, and in the right place ; for the discoverers are few, and their disciples can oftener receive than originate. But such instruction requires to be tempered by more methodical processes in the lower institutions. Benjamin Peirce, for instance, gave an enormous stimulus to mathematical thinking in this country ; but his methods of teaching were hardly adapted to the average Freshman and Sophomore.

GENERAL REMARKS.

Mathematics, as a branch of study, is very homogeneous. The difficulty, to a child of six, of understanding the processes of ordinary arithmetic, is very similar to that which the riper student finds in mastering the integral calculus. The reasoning is abstract in either case, and the combinations rather too numerous to be grasped at once. If you examine any college student who claims that he cannot, and never could, understand mathematics, you will find that some branches he understands very well indeed; about others, he will prove more indistinct. You will finally reach the point where he became wholly discouraged, and gave it up as hopeless. Many students cannot add quickly and correctly; the principle of *carrying* seems to trouble them, as is painfully evident in the average computations of novices. Fractions are often very indistinctly taught and learned; there are many that stumble on negative quantities, radicals, logarithms in algebra. In geometry, the measurement of the circle is a great rock of offence; and it is quite rare to find a pupil who sees distinctly and clearly the lines in space required in solid geometry.

Now, all this shows what is a great evil in our schools, — teaching by the inexperienced and inexpert. The pupil who cannot *carry* correctly, has never been taught the analysis (according to Grube) of the first hundred numbers.

$$\begin{array}{r} 539756 \\ 385429 \\ \hline 925185 \end{array}$$

Here I have set down, at random, two numbers : 56 and 29 are 85 ; not 9 and 6 are 15, set down 5, 1 to carry to 5 is 6, and 2 are 8. Again, 97 and 54 are 151 ; 53, 38, and 1 are 92.

This process, which I believe is like that of all skilled computers (except that they usually add from left to right), saves at once half the carrying, and the little which is left stands out in most cases visibly to consciousness as having some meaning, because the amount carried is far less than most of the other numbers added. But can the average primary teacher perform Grube's operations quickly and surely?

Now, in arithmetic, throughout, time must be taken away from riddles, puzzles, operations with enormous numbers, long sums on the slate or blackboard, and given to solid work in the elements of numbers, especially mental work, to extemporalia, to practical applications of arithmetic which really mean something.

This work is hard for the teacher, and requires strong, vigorous persons, who are well educated, for the lowest classes. And this again involves expense, which is often a bar to the best results. Let the great city superintendents and boards of education, then, fully realize their responsibilities, and keep themselves fresh and free from hampering traditions.

In the middle schools, the city grammar schools, the town high schools, the lower academies, etc., algebra is necessary, and geometry indispensable. Let the use of letters to express numbers and quantities, and the employment of the negative sign of quantity, and the geometrical forms and elementary relations be made the matter of very careful and thorough teaching, so that the correct conception and application of these things shall be instinctive with the vast majority of the pupils.

The teachers in these schools are more apt to be judged by "success in teaching" than anything else. They consequently are almost always anxious to improve. Superintendents and principals have in those grades a heavy responsibility. The danger is that they will be overworked; their offices, like those of all teachers in this country, and perhaps all others, are not fully appreciated; so that an ambitious and able superintendent or principal is liable to find himself unable to keep up his studies, and he is often thwarted in his plans and methods. The semi-political elections of teachers are a great evil.

The tradition is ingrained in us, of regarding the teacher's profession as a stepping-stone to something better, and so long as the teacher does not have leisure to study and power to apply his ideas in his teaching, just so long will he be discontented and anxious to get into a more independent profession.

As America becomes richer, the great benefactions to universities and colleges do something to raise the teaching craft. A bright young graduate who has made his mark as a teacher not unnaturally looks to Alma Mater to give him a position. If his work in lower schools is original and good, and he is promoted to the college, he will be missed lower down, and other like men be sought in his place. Thus the tendency to improve, if the college has it well developed, is propagated in the lower schools. But if colleges are backward, hold to antiquated and traditional methods, if their professors become sleepy from too easy positions, and lose their interest in higher study, their negative influence upon the lower schools will be very oppressive.

Fortunately, however, the colleges are fast getting out of their ruts. While the clamor for the repression of Greek is not a fair one, it reveals a popular belief that the colleges have been, in past times, unprogressive. But, Greek or no Greek, these institutions are thoroughly awakening to new methods. Just as soon as any college can get a large body of students who are non-Grecians and yet thoroughly trained, will there be a relaxation of the terms of admission on this point, or at least some method by which these men can receive a higher education. Whatever course of studies keeps unripe boys from premature graduation will always be unpopular with a large class of well-to-do parents. What every college faculty most desires is a good yearly accession of sound scholars ; and at the present da*y* there is a full practical acknowledgment that these young men are worth hard work on the teacher's part.

There is no department of college study in which improvements in method of instruction are not now making. The young men who return from Germany are not always well up in German pedagogy, it is true; but within a few years the older ones have become far more objective in their methods.

Laboratories, observatories, scientific collections, art museums, libraries, are fast advancing, and hardly any subject is now taught by text-book method only. And American scholars and scientists are becoming numerous enough to claim their rightful place in education; ignorance of the higher branches of a subject is now more rarely a positive recommendation to a teacher of the elements.

A "mathematical laboratory" is something not often mentioned; but I think the thing, however we name it, a necessity. It will contain, in part, such things as are ordinarily contained in a physical laboratory, but such as relate to ordinary, not purely scientific measures. Thus a pair of scales may be considered to belong to mathematical apparatus rather than physical, so far as weighing is a branch of arithmetic; a surveyor's chain, so far as the expansion by heat or the elasticity of the chain is not to be taken into account. If a good high school could exist, which taught mathematics and not physics, its laboratory would. contain all tools relating to the weights and measures needed in an ordinary community. -

At any rate, let not the teacher of mathematics imagine, if he does not teach physics, that he can without damage avoid all opportunities for laboratory practice on his pupils' part. Many young people do not know how to measure a room in feet and decimals, and get its area. So, too, careful weighing with good commercial scales or the steelyard is a useful arithmetical lesson.

PROGRAMME OF COURSE OF STUDY.

The ideal programme of study is about as follows. I give only the order of subjects, — in parallel columns : —

Primary Arithmetic.	Notions of Form, Drawing.
Arithmetic, through Rule of Three.	Rudiments of Geometry.
Universal Arithmetic. ⎫ Simple Equations. ⎭	Plane Geometry (one or two books).
Algebra, through Quadratics.	Plane Geometry.
Algebra, Series, etc.	⎧ Solid Geometry. ⎨ Conic Sections. ⎩ Plane Trigonometry.
Theory of Equations (begun).	Analytical Geometry (rudiments).
Calculus (rudiments).	Spherical Trigonometry.

Commercial arithmetic, for those who are to go into business, goes with algebra through quadratics, or replaces it. Of course this programme is somewhat variable ; but the main principle, that a course of arithmetic must run parallel with one of geometry from the beginning of a school course to the end, is one which is laid down by the best educators since Pestalozzi's time.

CONCLUSION.

It is likely that a large share of those who have read the previous chapters will think my suggestions unlikely to be realized, and that any one who has tried to get practical help from them will feel somewhat discouraged. To such I would say that educational theorists are often distasteful because they seem to require too much at once ; but the uprooting of old methods and the implantation of new ones, like the driving out of the Canaanites in the Holy Land, is apt to be a very gradual process.

The professor, superintendent, or teacher, is but one factor in the work of education. The programme of studies is usually fixed and badly arranged, and it is difficult to change it. When changes are made, however, they can usually be made in the right direction.

The young teacher fresh from college or the normal school will usually teach as he or she has been taught; if the college teaching is good, better than the average in the locality, improvements will gradually be made round about. And college teaching has, as I have said before, greatly improved of late years, especially in the objective and practical direction. Such a young instructor, if he is not overworked, can do much towards realizing a high ideal in his profession; and I trust the modern tendency towards a higher estimation of this profession will continue.

But such a young teacher, especially if but little leisure is allowed, is apt to overlook the need of studying the theory of teaching — the science of adaptation. He is so enthusiastic for the new truths he has learned in the higher school, that he does not see that his average pupils cannot at once come up to his standard. While admitting, at once it may be, that the true order of teaching is from the known to the unknown, he does not quite realize the "invincible ignorance" of the pupils; and is apt not to interest them, but to compel them to learn. Of course, compulsion to industry is a good thing; but many pupils lack industry for the very reason that learning has always been made hateful to them.

As a college professor I have known a great many students who suffered extremely and were a grief to all their friends, including their professors, simply from loose and disorderly habits of work. This, of course, is partly the result of heredity, partly of poor domestic training; but partly, I am very sure, from dogmatic instruction by inexperienced teachers. The young teacher is apt to impart, or try to impart, information, instead of producing good habits.   How is it with those who are more experienced?

I am inclined to think that our school systems, as a whole, put too much of the work upon the pupils and too little upon the teachers.   Not that the latter have, on the whole, less to do than those of foreign countries ; but in backward schools their labor is wasted by great subdivision, to accommodate the irregular progress of the pupils and the whims of parents ; while in the more advanced city schools too much time is often given to machinery and the marking system.   In certain cities the superintendent and the principals impress their methods too much upon their subordinates, who, for their part, have but little scholarship and individuality.   These evils are very closely connected with bad municipal government.

Experienced teachers, if they have leisure and disposition to improve, are always the best ; but youthful enthusiasm sometimes attains far better results than the extreme conservatism of old age.   A special difficulty in America has arisen from the enormous rapidity of progress ; the *laudatores temporis acti* still hold to a state of things which is intolerably old-fashioned.   In trigonometry, for instance, we have text-books whose stereotype plates are not yet worn out, but whose ideas are those of a past century ; and it is extremely difficult to modernize a pupil taught with their help.   The student must, to solve his triangle, draw a figure and make a proportion ; he has to be trained to work from formulæ.

Many such teachers are skeptical about the necessity or usefulness of studying educational theory.   At bottom, this refusal or neglect amounts to the affirmation that tradition and the general progress of the age, combined with natural ability to teach, are a sufficient guide to good teaching ; but it often happens that such an instructor repels the effect upon him of the world's progress, and becomes a conservative pure and simple.   Such a course is very detrimental, and is a serious drawback to the benefit to be derived from his experience ; his classes are apt to work mechanically and without interest. The best thing for an experienced teacher to do is to progress slowly but constantly in the improvement of his methods. The books he must study are, in mathematics, those of a scientific character in the branches he teaches, and good text-books in the portions just beyond his course ; also those which treat of the applications of what he is teaching, and finally works on pedagogy. He will do well in these latter studies to take up practical books, not theoretical ones only.

In conclusion, I may be permitted to say again, what I have said before, that the mathematics which is serviceable in general education is not that of tradition and the English universities of half a century or more ago, but the practical form of the science first developed on the continent of Europe. This is the form which our colleges and technical schools need, and which can be made intelligible and interesting. Much of what has been taught in our schools must and will be dropped as artificial and ugly, to be replaced by that which is natural, beautiful, and useful. American mathematicians will do well to go on in the lines traced for them by Benjamin Peirce and Chauvenet ; and our teachers will need to study this science, and at the same time learn the great art of Pestalozzi and his followers.

APPENDIX.

THE kindly critic of this little work in the *Nation* of February 24 has suggested that Grube is not well known. I find his name mentioned in Professor Hall's Bibliography, and the title of his book is there given. Since writing the monograph I have obtained a copy of the book, "Leitfaden für das Rechnen in der Elementarschule," 6te Auflage, Berlin, 1881 ; about 150 pages octavo. The price in Germany is 1 mark, 80 pfennigs, or about 45 cents. Soldan's abstract is published in this country by the Interstate Publishing Company.

Grube's leading idea is that the separate numbers up to 100 must be studied as individual objects ; not, of course, by the use of counters beyond the first few exercises, and at proper intervals later ; but taking, as I have pointed out on page 20, each number as a thing to be analyzed by the pupils. When the class has arrived at 89 in the course of work, as the 89th step, they are already familiar with the qualities of the preceding numbers by using proper synthetic exercises and problems.

In explaining the steps to be taken with the first hundred numbers, the author occupies as much space as is contained in the whole of this monograph ; all of it is very instructive.

Grube takes the ground that the systematic study of numbers, according to rules or methods of treatment, such as numeration, notation, addition, etc., must be preceded by at least two years' work in the numerical elements ; in order to train the mind to deal with numbers systematically, it must already be familiar with the numbers themselves. The usual practice, so far as it attempts to teach things and their systematic arrangement at the same time, is wrong; the true method of teaching a system is to use as its material that which is already well known.

I may be permitted here to say that I have, and have had for many years, an almost unconscious habit of analyzing numbers which casually present themselves, — such as the street numbers of shops and houses. A distinguished Harvard professor of science, formerly tutor in mathematics, told me the same thing about himself with regard to the numbers of the hymns as set up in church; and I have no doubt my friend and critic can detect the same thing in his own mental constitution.

Those who wish to improve their methods of teaching mathematics will do well to study the excellent manuals printed, or published, for the use of elementary schools; a good one is Mr. John T. Prince's "Courses and Methods" (Boston, Ginn & Co.); as well as other pedagogical books. Much which I should need to say to amplify this monograph, as the critic desires, is merely the ordinary working-out of common school methods to be found in these works.

The facts which are brought out in the *Nation* (as before quoted) about lack of elementary training in such subjects as the meaning of "seven per cent," "three-thirteenths," and "two right angles" simply prove that the pupils have not had these *objects* (using the word in its mental sense) properly presented and impressed upon their minds.

For example, the idea of one-half, the simplest of all fractions, is, according to Grube, the subject of the first step of the course for the fourth year; then, when the pupil is about nine years old. Grube occupies rather more than two pages in explaining how to teach it, and wishes it to be followed by the "thirds," the "fourths," and so on. But in dealing with the "thirds" the sixths come naturally by subtraction from the half; so that when the "sixths" are nominally reached, the pupil is quite ready to deal with pretty large denominators, up to thirty at least. Grube recommends that for this first part of the course in fractions half a year should be taken; thinking, quite justly, that after the pupil can analyze and combine by addition and subtraction every fraction with denominator six or less, he is quite ready to go on rapidly in a systematic course of fractions according to the four ground rules.

Grube's method for arithmetic seems to me that which best prepares the mind for systematic work by varied exercises of all kinds in the rudiments. It is partially employed in the late edition of Warren Colburn's Arithmetic, the editors of which, as I understand, were two distinguished Cambridge mathematicians. They do not give the analysis, according to Grube, beyond twenty, so that a teacher would need to study Grube's book in addition to employing Colburn's.

Let me say, in concluding this short appendix, that I find that on page 30 I have been misled into expressing myself ambiguously about the "facts" of geometry. My opinion is that in that matter I differ from Dr. Hill. The "rudimentary geometry" I wish to see taught in the lower schools has been systematically worked out by a great many German writers; it enables pupils to *geometrize* without, perhaps, the ability to put their demonstrations into words as demonstrations; just as in arithmetic one multiplies both ways, and soon becomes aware that 9 times 7 equals 7 times 9 without the ability to prove that $ab = ba$.

Mere facts in geometry are, perhaps, not fruitful at all; but the theory of instruction which I have been led to accept lays great stress upon the separation of difficulties for the child's mind into various classes, according to the maturity which they demand of the pupil, and upon the preparation of children six to nine years old for later systematic instruction by a half-playful, at least an objective, presentation of the rudiments.

Several indications have reached me that the higher teachers in this country are not all well aware of the vast progress which the didactic art is making in the primary schools, or if aware that something is going on, do not appreciate it. In studying pedagogy I have continually derived great benefit from familiarity with normal school work and the maxims and practice of the best primary teachers.

SCIENCE.

For descriptions of forthcoming books in Science see announcements at end of this catalogue.

Organic Chemistry:

An Introduction to the Study of the Compounds of Carbon. By IRA REMSEN, Professor of Chemistry, Johns Hopkins University, Baltimore. 374 pages. Cloth. Price by mail, $1.30. Introduction price, $1.20.

THIS book is strictly an introduction to the study of the compounds of carbon, adapted to the use of scientific schools, medical schools, schools of technology, and colleges. It takes nothing for granted except an elementary knowledge of general chemistry. Special care has been taken to select for treatment such compounds as will best serve to make clear the fundamental principles of the subject. General relations as illustrated by special cases are discussed rather more fully than is customary in books of the same size; and, on the other hand, the number of compounds taken up is smaller than usual, though all which are of real importance to the beginner are treated with some degree of fulness. There is thus less danger of confusion than when a larger number is brought to the attention of the student.

Eleven editions of this book have been sold for the English market. It has been translated into German, Italian, and Russian. These facts, together with the estimates of professors of chemistry which will be found in our special circular, show that the book has already taken high rank on both sides of the Atlantic.

M. M. Pattison Muir, *Prof. of Chemistry, Cambridge Univ., England:* A great many students in this university are using the book; indeed, I think it will soon be used in preference to any other elementary work on the subject.

D. Mendeleeff, *Prof. of Chemistry in the Univ. of St. Petersburg, Russia:* The subject is treated in a manner so novel and so instructive that I look forward with confidence to seeing the book translated into other languages.

I

The Elements of Inorganic Chemistry:

Descriptive and Qualitative. A Text-Book for Beginners, based on Experimental and Inductive Methods. By J. H. SHEPARD, Professor of Chemistry, South Dakota Agricultural College, and Chemist to the U. S. Experimental Station, S.D. 397 pages. Cloth. Price by mail, $1.25. Introduction price, $1.12.

IT is a practical embodiment of **the modern spirit of investigation.** It places the student in the position of an investigator, and calls into play mental faculties that are too often wholly neglected. It leads him *to experiment, to observe, to think, to originate.* Coming as it does from the working laboratory of a practical instructor, who has had the constant advice of fellow-teachers in all parts of the country, this text may be fairly taken as an exponent of the latest methods of teaching chemistry.

Its distinctive features are: experimental and inductive methods; the union of descriptive and qualitative chemistry, thus allowing these kindred branches to supplement and illustrate each other; a practical course of laboratory work illustrating the general principles and their application; a fair presentation of chemical theories, and a conciseness which confines the work to the required limits.

The book covers more fully than any other the outline of work laid down in the report of the Conference on Chemistry to the Committee of Ten.

Our special circular on this book contains a list of over three hundred and fifty Colleges and Schools into which it made its way by merit alone, in a little more than a year's time, and also gives a large number of letters from teachers of the subject showing with what ease, profit, and satisfaction it is used in the class-room.

C. A. Schæffer, *Prof. of Chemistry, Cornell Univ.:* It is excellent. The plan is well conceived, and embodies the method by all means the best.

C. F. Chandler, *Prof. of Chemistry, Columbia Univ.:* An excellent book. A great degree of accuracy characterizes the entire work.

Otis C. Johnson, *Prof. of Applied Chemistry, Univ. of Michigan:* I like it so well I have nothing to criticise.

Ira Remsen, *Prof. of Chemistry, Johns Hopkins Univ.:* The book teaches that chemistry can only be learned in the laboratory, and that the book is only a guide to experimental work, — a lesson which certainly needs to be taught above all others in chemistry.

Herbert E. Smith, *Prof. of Chemistry, Yale Coll.:* A class working through it would gain an unusually good knowledge of the subject.

4 *SCIENCE.*

The Laboratory Note-Book.

For Students using any Chemistry. Board covers. Cloth back. 192 pages. Price by mail, 40 cents. Introduction price, 35 cents.

IT contains blanks for experiments ; blank tables for the reactions of the different metallic salts ; pages for miscellaneous matter ; and an extra chart for the natural classification of the elements, similar to that on page 221 of Shepard's Chemistry, which may be rolled into a cylinder by the student.

The advantages of using this note-book are, briefly, these : It saves time for the student and gives the teacher control of his work ; its size is convenient ; it is cheaper than an ordinary blank-book ; the paper is such that it readily takes ink without blotting or smearing, or it may be used with a lead pencil.

The value of systematic note-taking by the the student in chemistry can hardly be over-estimated.

Our Chemistry Circular contains fac-similes of three pages, prepared by the students in the Ypsilanti high school, showing how the book is to be used.

Geo. F. Sawyer, *Dept. of Chemistry, Martha's Vineyard Summer School:* It is the best in plan I have seen.

D. W. Batson, *Pres. Ky. Wesleyan Univ., Millersburg:* It is answering our purpose excellently.

Chemical Problems :

Adapted to High Schools and Colleges. By JOSEPH P. GRABFIELD and T. S. BURNS, Instructors in General Chemistry in the Mass. Inst. of Technology. Paper. 96 pages. Price by mail, 30 cts. Introduction price, 25 cents.

THIS book comprises the principles of stoichiometry, with separate chapters upon atomic and molecular weight, determinations and specific gravity of gases, and upon the principles of thermo-chemistry and its application to inorganic chemistry. The authors have added, as a set of general questions, a series of problems and reactions and the examination papers in general chemistry given at the Institute of Technology during the last ten years. The book contains no formulæ, students being led to solve the problems from the principles involved.

A Short History of Chemistry.

By F. P. VENABLE, Professor of Chemistry in the University of North Carolina. 171 pages. Cloth: Introduction price, $1.00. By mail, $1.10.

THERE has long been a need for a short, systematically arranged History of Chemistry written in the English language. This book is issued with a view to filling this want. It is concise, and yet full enough to give a connected view of the growth and development of the science. Stress has been especially laid upon the rise of theories and the discussions over them, the overthrow of the false and the survival of the true.

To the student, even the beginner in the science, as Victor Meyer has pointed out, the historical method is the most attractive and the most promising in results.

This book is the outcome of a course of lectures, tested during three years' use with students. All of the chief authorities upon the subject have been consulted, and, where possible, the original writings referred to. The aim has been to so digest and systematize this mass of material as to render it available for those desiring a general knowledge of the subject.

The book should be in every school and college library and is especially adapted for use as complement to the ordinary college courses in general chemistry.

A prospectus giving preface and table of contents sent free to any one desiring to see it.

The following are a few of the many commendatory words that have come to us concerning Dr. Venable's book :—

Ira Remsen, *Professor of Chemistry, Johns Hopkins University:* It is short and clear, and will, I think, prove of value to students and general readers. There is no book like it in English.

J. O. Reed, *Asst. Professor of Physics, University of Michigan :* The appearance of the book is timely and it contains much valuable matter in small space.

The Educational Times, *London, Eng.:* We strongly recommend this charming little book to teachers of chemistry. It is attractively written, and we have failed to find any inaccuracy. The author has fully justified his assertion that the study of the history of a science is one of the best aids to its intelligent comprehension.

A Laboratory Manual in Organic Chemistry.

Containing directions for a course of experiments, systematically arranged to accompany Remsen's Organic Chemistry. By W. R. ORNDORFF, Assistant Professor of Chemistry in Cornell University. 192 pages.   Boards.   Introduction price, 35 cents.   By mail, 40 cents.

INTENDED to be used as a laboratory guide in the study of the Compounds of Carbon, or Organic Chemistry, and to supplement the lectures and text-book work in the subject.   Careful directions are given for about eighty experiments on the derivatives of the most important series of hydrocarbons.   The student is constantly questioned as to the meaning of his work and the results obtained, and is led to deduce his own conclusions as far as possible.

The book has been in use for the past six years in the Organic Laboratory of Cornell University, and has given general satisfaction.   It has now been corrected and revised in the light of the experience gained in the past, and in the belief that it will prove useful in other laboratories.

A Laboratory Guide in General Chemistry.

For a twenty weeks course. By GEORGE WILLARD BENTON, Instructor in Chemistry, Indianapolis High School. 165 pages. Boards. Introduction price, 35 cents. By mail, 40 cents.

THIS manual is intended for the use of the student in working laboratories of high school grade.   It contains careful and detailed instructions for the performance of 156 experiments, an abundance of queries and blank pages for notes.   The experiments have been selected with a view to simplicity in apparatus and manipulation, and a high percentage of success at the hands of the average pupil.

As a foundation for quiz and recitation, and for the development of the general principles of inorganic chemistry, the course will be found suggestive and pedagogical.   It is adapted to any text-book.

Teachers will appreciate the attention given to details, which are so essential to beginners.   This feature of the book, having been especially considered as the course has grown up in actual laboratory practice, is most carefully and successfully carried out.

The table of References by Experiments as given in the appendix will be found a new and valuable feature.

The Elements of Chemical Arithmetic,

with a Short System of Elementary Qualitative Analysis. By J. MILNOR COIT, Master in St. Paul's School, Concord, N. H. 95 pages. Cloth. Price by mail, 55 cents. Introduction price, 50 cents.

THIS manual is divided into two parts. Part I contains the more important rules and principles of Stoichiometry, put in a very simple way to suit the comprehension of the average student. These rules and principles find application in a series of carefully selected problems which follow. Part II contains an elementary system of qualitative analysis intended for beginners. The most available and satisfactory tests for the more important metals and acids are given with simple tables of separation. Besides these are some carefully compiled tables for reference. The book will be found valuable as a companion for some regular text of descriptive chemistry or where it is desired to give an elementary course in practical chemistry. It has just been carefully revised and some valuable tables added.

T. H. Norton, *Prof. of Chemistry, Univ. of Cincinnati, O.:* It is admirably written, and well adapted to supplement the ordinary descriptive text-book or series of lectures preparatory to scientific courses.

Robert Peter, *Prof. of Chemistry, Kentucky State Univ., Lexington:* I can recommend it as a good, compendious, and cheap guide for Laboratory practice in Qualitative Analysis.

The Development and Present Aspects of

Stereo-Chemistry. By CHARLOTTE F. ROBERTS. Professor of Chemistry in Wellesley College, Mass. 160 pages. Cloth. Introduction price, $1.00. By mail, $1.10.

THE object of this little book is to present in a somewhat elementary and compact form the most important results of the stereo-chemical investigation of the past twenty years. It is hoped that it may give to students of chemistry, who have not had the time or opportunity to follow the original papers which have appeared from time to time, a general idea of the work which has been accomplished, and of that which remains to be done. Since constant references are made to the original papers, the book may well be taken as the starting-point for a more extended study of the subject.

The principal topics are: 1st, the general principles of stereo-chemistry; 2d, the work done, principally by Von Baeyer, in connection with closed rings; 3d, the stereo-chemistry of nitrogen; 4th, light thrown on theoretical problems by stereo-chemical investigation.

First Book in Geology.

By N. S. SHALER, Professor of Paleontology, Harvard University. 272 pages, with 130 figures in the text. Cloth: introduction price, $1.00; price by mail, $1.10. Boards: introduction price, 60 cts.; by mail, 70 cts. Teacher's Manual. Paper, 74 pages. Retail price, 25 cts.

DESIGNED to give public school pupils and general readers a few clear, well-selected facts as a key to the knowledge of the earth. The aim is to illustrate the principles of geology by reference to as many facts of familiar experience as possible, such as; Pebbles, Sand, and Clay. The Making of Rocks. The Work of Water and Air. The Depths of the Earth. Irregularities of the Earth. Origin of Valleys and Lakes. Movements of the Earth's Surface. Place of Animated Beings in the World. Sketch of the Earth's Organic Life. Nature and Teaching of Fossils. Origin of Organic Life, and a Brief Account of the Succession of Events on the Earth's Surface, etc.

The Teacher's Manual contains seventy-four pages of directions for those who use the book in class instruction. The average reader who desires to get a glance at geology and a general notion of its bearings on ordinary life will find this edition of exceeding interest.

The special circular on this book shows how it has been received by teachers of Geology. The book has been translated into Polish, and is soon to be translated into German. It is also being used in many schools as a Supplementary Reader, the edition in board covers being especially prepared for this purpose.

John C. Branner, *Prof. of Geology, Stanford Univ.:* With a view to urging the use of some elementary book on geology in the schools of this State, I have examined Prof. Shaler's Geology. I cannot do better than recommend it.

F. W. Hooper, *Teacher of Nat. Science, Adelphi Acad., Brooklyn:* Of all the attempts at making elementary text-books on the natural history sciences, this is the most successful.

N. A. Cobb, *Teacher of Natural Science, Williston Sem., Mass.:* I un-hesitatingly say that it is the best text-book on geology for young people that I have ever seen.

Alexander Winchell, *late Prof. Geology, Univ. of Mich.:* Its marked departure from the methods of the old didactic treatises which have done so much to put geology at a disadvantage, in comparison with botany, is in the direction of common sense and in the interests of science and education. I believe the book will have, as it deserves, an extensive use in the school.

Object Lessons and How to Give Them.

In two volumes. Manuals for Primary and Intermediate Teachers. By GEORGE
RICKS, Inspector of Schools, London School Board. First volume, 200 pages.
Second volume, 212 pages. Retail price, 90 cents each.

THESE two graded volumes are by the author of "Natural History
Object Lessons," which is so popular and so widely used, and
are designed to aid teachers in the lower grades in the same way as
the more advanced book gives aid for higher grades.

Two points of fundamental importance are considered: (1) the selection and adaptation of lessons to the capacities of children; (2) the
method of giving the individual lesson. By means of the complete
system here given, lessons are graduated to the mental conditions and
previous training of pupils. Each lesson, while fulfilling its own
special purpose, forms a link in a chain. The method of conducting
the lesson is such that the child is not allowed to be a mere passive
recipient of information, but is led to exercise his own mental powers.
The lessons are suggestive rather than exhaustive; some are little more
than outlines; a few are worked out more fully as models for young
teachers. The experiments are numerous and interesting, yet simple
and inexpensive. The contents of the books are as follows: —

FIRST VOLUME. · Chapters I and II train to habits of observation
and comparison, rather than impart information. III shows by means
of experiments special properties upon which uses of familiar objects
depend. IV imparts useful information about common necessaries of life.
V completes course of "Systematic Object Lessons" for infant schools.

SECOND VOLUME. Chapter I deals with common properties of
solids. II deals with similar properties of liquids. III, IV, and V
bring the reasoning faculties more fully into play.

W. D. Sheldon, *Vice-Pres. of Girard
Coll., Phila.* I have examined them with
a good deal of care. The lessons are admirably arranged and developed. I shall
recommend them.

A. C. Boyden, *Teacher of Science,
Normal School, Bridgewater, Mass.:*
They are excellent books and I strongly ad-
vise you to put them on this market.

Geo. L. Chandler, *Supv. of Science,
Public Schools, Newton, Mass.:* I think
Part II (which I have examined) a book
that will find a place on the teacher's
study table for help in preparing science
work.

F. H. Bailey, *formerly Teacher of Sciences, Mrs. Shaw's School, Boston:* They
are invaluable. I cannot spare them a
day from my work.

Elementary Course in Practical Zoology.

By B. P. COLTON, A.M., Prin. of High School, Ottawa, Ill. Cloth. 196 pages. Price by mail, 85 cents. Introduction price, 80 cents.

A BOOK to lie open before the student while he is studying the animal itself. It first tells him where to find his specimens ; how to observe their habits and habitats; in a few cases their metamorphoses and modes of development ; how to collect and preserve ; and, finally, how to dissect them. In short, it is a guide to the study of animals rather than a mere descriptive zoölogy.

A series of typical representations of all the animal sub-kingdoms is represented ; after the study of each, the student is led to compare them one with another, noting their resemblances and differences ; thus he learns how to classify animals instead of memorizing a system of classification.

This method gives the student a better view of the animal kingdom than any amount of mere reading, and, at the same time, develops and disciplines his powers of observation and description.

The special circular on this book gives a large number of strong commendations ; a list of places that at once adopted it ; and an excellent article on the study of Zoölogy, written for " The Dial" by Carl H. Eigenmann, Dept. of Zoölogy, Ind. State Univ.

David S. Jordan, *Pres. of Leland Stanford Jr., University, Cal.:* I regard it as the only text-book in general zoölogy yet published which is *fit to be used* in high school classes. It is on the right plan, and the method is admirably carried out.

Nature, *London :* In consideration of the pernicious rubbish which, even yet, occasionally finds its way into our own elementary schools under the guise of the elementary text-book of Science, it is pleasant to reflect upon the merits of this work.

Prof. Alpheus Hyatt, *Boston Society of Natural History:* Mr Colton's desire is to cultivate the faculty of observation, and I think he has succeeded in producing a very valuable help to the teacher of manual science.

T. C. Mendenhall, *Pres. Polytechnic Inst., Worcester, Mass.:* I am delighted with the book. I think the author has struck " pay-rock " for the department of science.

R. Ellsworth Call, *Manual Training School, Louisville, Ky.:* We have had 135 copies in use. Enthusiasm has been awakened by the actual study of the objects,—an enthusiasm which has been directed to definite ends by this little work. The book deserves most abundant success, and I believe will receive it.

Albert H. Tuttle, *Prof. of Zoölogy, Ohio State Univ.:* It is the most useful book of its kind ever published in this country, not only for high school use, but for beginners in zoölogy in colleges as well.

An Astronomical Lantern.

Invented by Rev. JAMES FREEMAN CLARKE, Boston. The face (6 1-2 by 10 inches in size) is of ground glass, behind which lights may be placed. Thirty-two constellations are photographed upon seventeen slides of semi-transparent card board, and stars of four magnitudes are represented by perforations of proper size. Price with the slides and a copy of " How to Find the Stars," (see below) $4.50. The whole carefully packed in a wooden box, with sliding cover.

THIS useful piece of apparatus is designed to facilitate the study of stellar astronomy. It is intended for beginners in astronomy in schools and in families, and for all, both old and young, who desire to become acquainted with the constellations.

The student wishing to observe any particular constellation or cluster has only to light a candle within the lantern, insert the appropriate slide, and go out into the night. Holding up the lantern in one hand, he can compare the constellation as it appears on the lantern with that in the sky, until he becomes perfectly familiar with the latter.

C. A. Young, *Prof. of Astronomy, Princeton Coll.:* I find it to be an admirably contrived apparatus for its purpose,—simple, easily managed, and effective. I think an adequate knowledge of the constellations could be obtained by its use, in connection with the little book that accompanies it, more rapidly and easily than from the most elaborate and expensive celestial globe.

C. S. Lyman, *Prof. of Astronomy, Yale Univ.:* I have never known any contrivance that could compare with this Lantern for saving alike time, trouble, and eye-sight, and rendering such study attractive and easy.

O. C. Wendell, *Harvard Coll. Observatory:* It combines in a high degree simplicity with clearness, and, for beginners and amateurs, I think it has no equal. I am confident that the names of the principal stars and constellations can be learned from it in half the time required to learn them from an ordinary map.

How to Find the Stars.

By Rev. JAMES FREEMAN CLARKE. Paper. 47 pages. Price, 15 cents.

THE object of this book is to help the beginner to become better acquainted with the visible starry heavens; to know the winter and summer constellations and the principal fixed stars. It shows the position of the constellations at different periods of the year. The most interesting objects at each period of the year, especially such as can be found with a telescope of moderate power, are indicated. A description of the Astronomical Lantern (mentioned above) is appended.

www.ingramcontent.com/pod-product-compliance
Lightning Source LLC
Chambersburg PA
CBHW022028080426
42733CB00007B/768